Prospective Landscapes: A Retrospective

Drawing Exhibition 'Vortex'

Landscape Architect, Ryoko Ueyama,
Colleagues and Collaborators

天と地をつなぐランドスケープ 渦展

上山良子と渦なかま

Ryoko UEYAMA
Between the Sky and the Earth: The last flight of the Hot-Air Balloon 'Ubiquitous'

Ten thousand meters above northern Europe, I encountered the aurora floating in the sky over the abyss. A desert sunset bathing the sky in crimson. Wildly winding rivers on the prairie. Dancing clouds over the tropical seas. The landscape of NON-DESIGN, which exists infinitely and limitlessly, attracts us. From it, we gain a moment of peace in the majestic nature and casual scenery that we encounter. These experiences were the everyday scenes on board SAS that I have chosen to admire the Scandinavian design at that time.

A few decades have passed since that flight to NON-DESIGN on the hot air balloon 'Ubiquitous.' As a creator of the landscape, I have carved a new layer on the earth, and now I am convinced that my ultimate ideal of life is 'ubiquitous', a beautiful word that means "to exist here and there at the same time". It transcends time and space. What a wonderful thing! It is the "philosopher's stone" that mankind has sought for many years. "The 'philosopher's stone' is the sublime harmony, transcendence over every conflict, absolute liberty, and deliverance from time and causality" (Rudolf Bernoulli). Cezanne, who stated that "nature is infinite", learned much from nature and expressed it on canvas. In the same sense, landscape architects work on large-scale projects that transform the infinite nature into human environment with nature as materials. Their task is to 'design with nature'. The possibilities of landscape design are endless.

Over 20 years have passed since the quest for possibilities of landscape design *, and now, let's take a journey aboard the hot air balloon 'Ubiquitous' once again. This final flight is a trip around a new layer of earth that we have carved.
Creators have a mission to continuously create such a space where people can stop and "think about themselves". It is a journey of endless joy of creation beyond time and space.
The unreliable movement of the hot air balloon is perfect for our purpose. This time, the flight will be accompanied by birds (eye from the sky), dogs (eye on the surface), moles (eye from the underground), and children. We want children to enter a new era where they can experience the starting point of creating a new place!

Let's start the journey to the 'Archives Island' from the 'UZU: Vortex' tour. 'Vortex' has led me to the optimal solution to carve the earth. In the process of archiving my works, I have rediscovered that many of my ideas have been originated from 'UZU'. So, first, enjoy the world of 'UZU', then go to the 'Archives Island'!
Now, let us drop the sandbag of prejudice onto the ground. This marks the departure of the 'Ubiquitous' as it starts its journey towards the history of coexistence between humanity and nature, and the paradise that lays ahead in the future.

* Ryoko UEYAMA, The journey of "Hot Air Balloon UBIQUITOUS"-In search for the possibility of landscape, "icon" Vol. 0 (1986/10) to Vol. 6 (1987/7).

上山良子
空と地球の間にて：熱気球Ubiquitous号の最終フライト

深淵に広がる青黒い空に輝くオーロラと出会った北欧の上空1万メートル。深紅に燃える砂漠の落日。けだるくうねる大草原の川。熱帯の海上の雲の乱舞。無限に存在する人間の手でデザインされていない風景、いわば、NON DESIGNの風景は限りなく我々を魅了する。雄大な自然の中に、そして又ふっと出会うさりげない風景の中につかの間の安らぎを見出す。これらは北欧に憧れて選んだSASの機上での日常の風景だった。

それから、何十年の年月が流れたことか。風景の創り手として、大地に新しいレイヤーを刻んできた今、自分の究極の生き方の理想が、ubiquitous—「同時にあちらこちらに存在する」「遍在する」だったのだと確信している。時間と空間を超越して、そこここに出没する。何と素晴らしいことか。ubiquitous、それは人類が長年の間求めてきた「賢者の石」、すなわち「すべての対立を超越した崇高な調和、絶対の自由、時間、因果律からの解脱としての賢者の石」（ルドルフ・ベルヌーリ『錬金術』種村季弘訳）だ。
「自然は無限だ」といったセザンヌは自然から多くを学び、キャンバスに表現した。ランドスケープアーキテクトは、無限の自然を組み替え、自然を素材にして表現する大事業に取り組む宿命を負っている。
ランドスケープデザインの可能性は無限だ。

ランドスケープの可能性を求める旅*から20年以上の月日が経った今、再び、「熱気球 Ubiquitous号」に乗って旅に出よう。この最終フライトは私たちの手で刻んできた大地の新たなレイヤーを巡る旅になる。人々がその場で立ち止まり「自分自身を考える」そんな空間をクリエーターたちは創り続けていく使命がある。その旅こそ、時空を超えた限りない創造の歓喜の旅なのだ。
熱気球の頼りなげな動きは、私たちの目的にピッタリだ。そして、このフライトには、鳥（空からの眼）と犬（地表からの眼）ともぐら（地中からの眼）、そしてこどもたちも一緒につれていこう。新しい時代を担うこどもたちには、場づくりの原点を体感してもらいたいから！

「アーカイヴス島」に向かう旅は、「うず：渦」巡りからはじめよう。大地を穿つ最適解を導いてくれた「うず」。今回、アーカイヴスをまとめる過程で、私の発想の原点が「うず」から発していることを再発見したからだ。
まず、「うず」の世界に遊んでから「アーカイヴス島」へ！
さあ、偏見という砂袋を地上に落とそう。熱気球 Ubiquitous号の出発だ！人類と自然との共存の歴史の彼方へ、また未来の楽園に向けて。

*上山良子、熱気球UBIQUITOUS号の旅—ランドスケープの可能性を求めて、『icon/イコン』Vol.0（1986/10）からVol.6(1987/7)に連載。

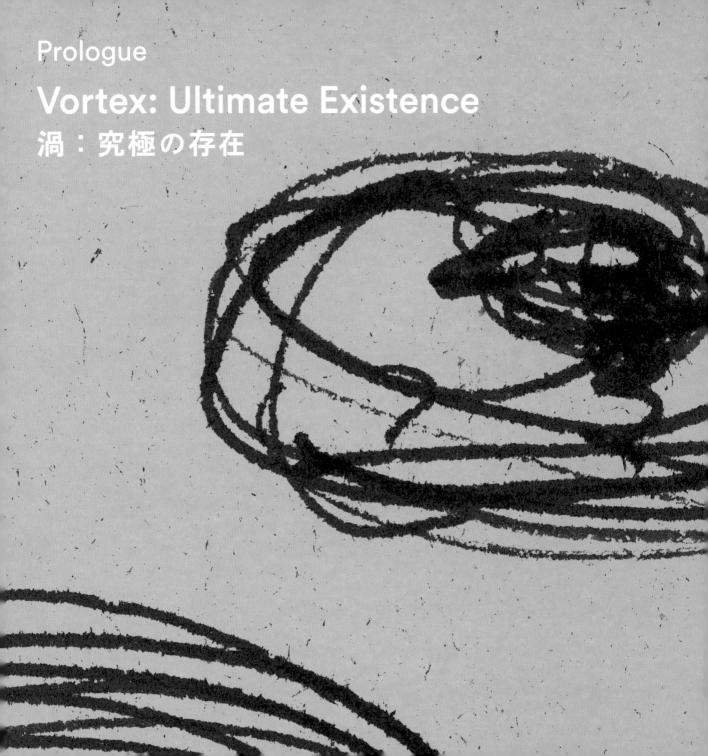

Prologue
Vortex: Ultimate Existence
渦：究極の存在

Let's find 'Vortex'!

In the dark winter, North end of the Earth.
Look! Huge 'vortex' of Ice! It's a big skate rink!

北のはてにある国のま冬。
凍てつくさむさの中、銀色の氷の'うず'をみつけた。大きなスケートリンクだ！

Vortex fountain at the midnight sun in the summer.
夏の白夜、'うず'は噴水になる。

Töölönlahti Lake Landscape Design International Competition, Helsinki; Invited Participation: The Purchased Prize, 1997

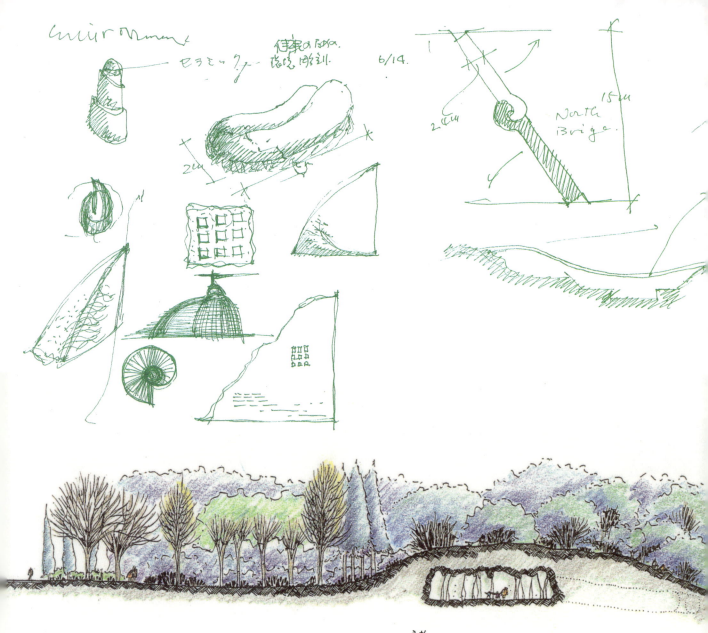

Draw the ideas by hand!
Ask to the God of the earth,
what this place wants to be!

アイディアを描こう！
ここはどうなりたいのか、
大地の神さまにきいて！

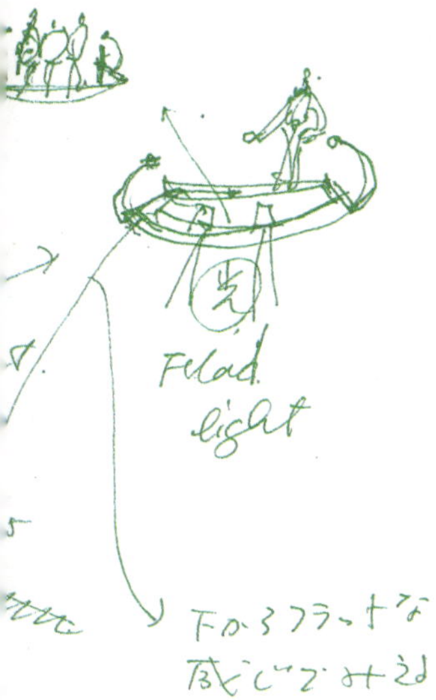

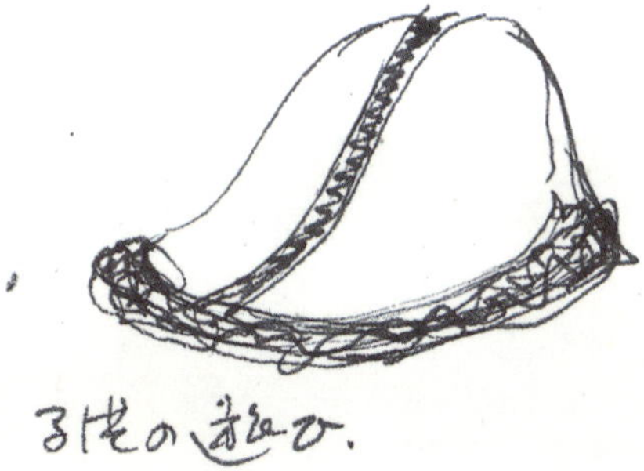

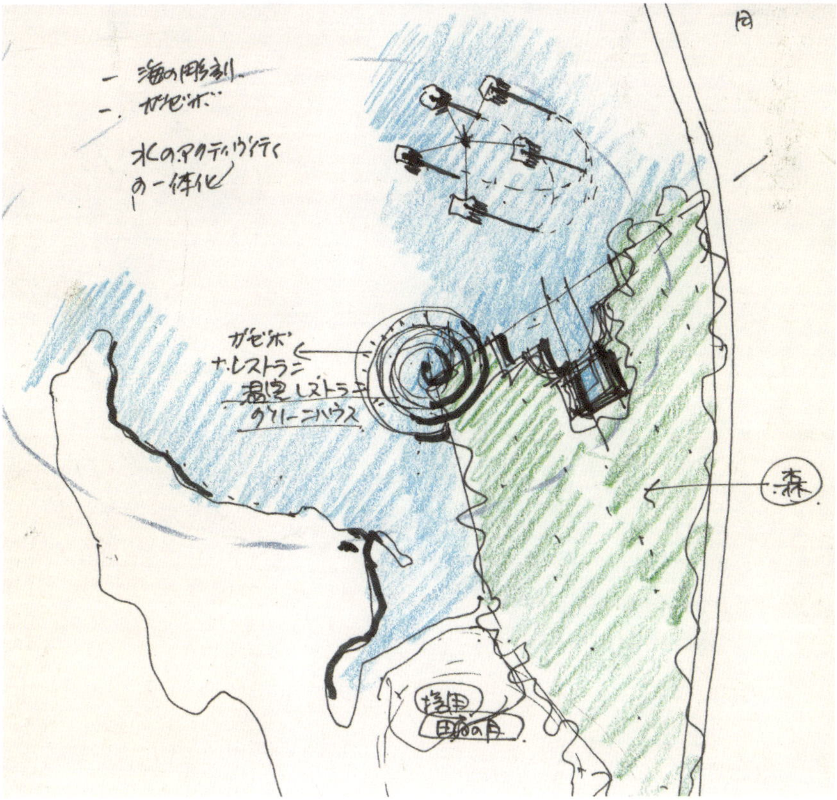

Töölonlahti Lake Landscape Design International Competition, Helsinki, Invited Participation: The Purchased Prize, 1997 FA2230000109 / T29000102

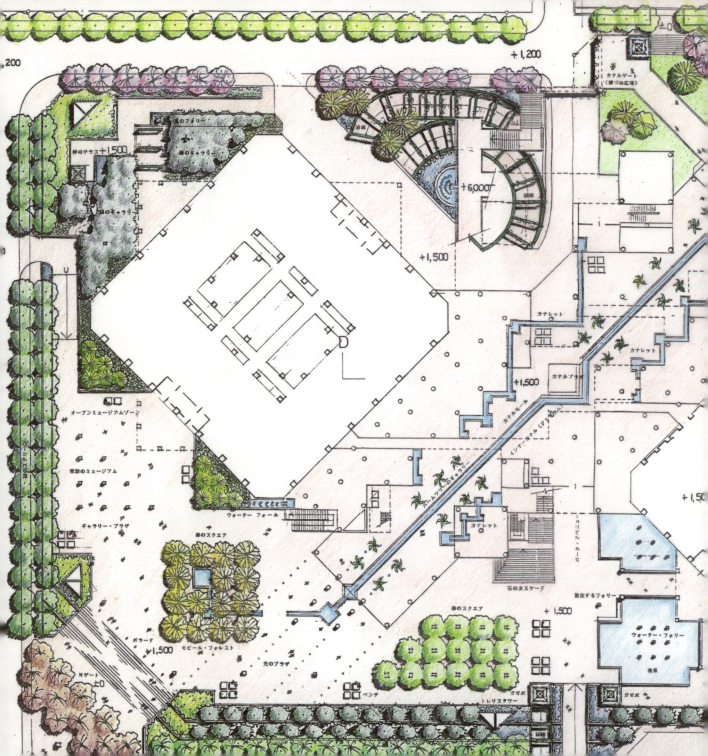

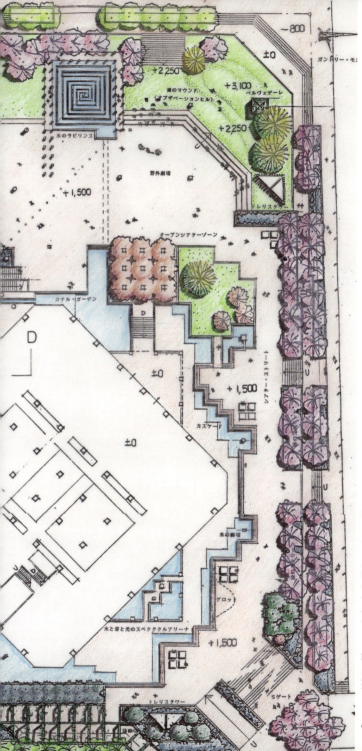

**Flying over the Tokyo Bay,
we can see an square 'vortex'.
The Water Theater!**

東京湾の上を飛んでいると、
四角い'うず'を見つけた。
あれは、水の劇場！

Seavans: Water Theater: Labyrinth, Tokyo, 1987   T02000101

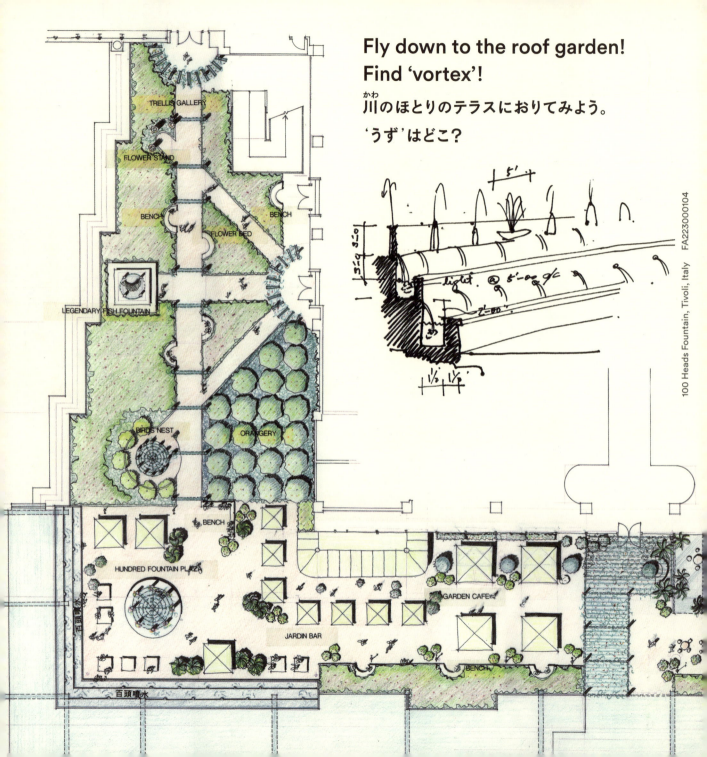

TTSC Tamagawa Takashimaya SC 20th anniversary renovation, Tokyo, 1988

100 Heads Fountain! Ammonites and Fishes mean the memory of the site.

たくさんのアンモナイトから水が噴き出す。
川の近くは、魚や貝の出番だ。

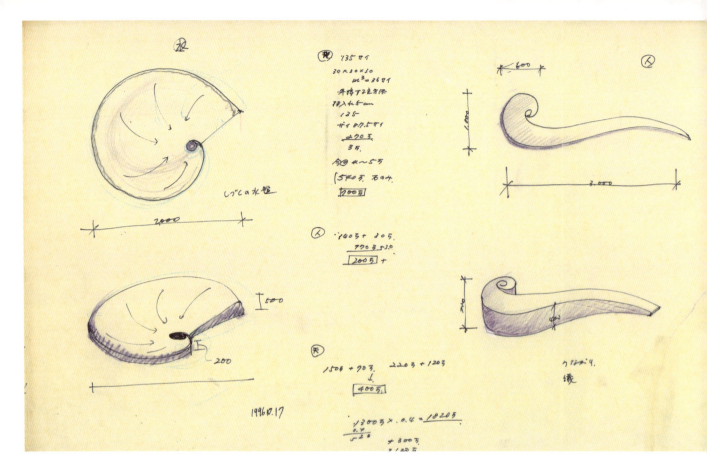

In spring of the snow country,
people can rest on those stone sculptures.
Heaven, Earth, Person, Wind and Water.

雪国の春、山を登る人たちがひとやすみするところ。
天、地、人、風、水。石の彫刻にねっころがろう。

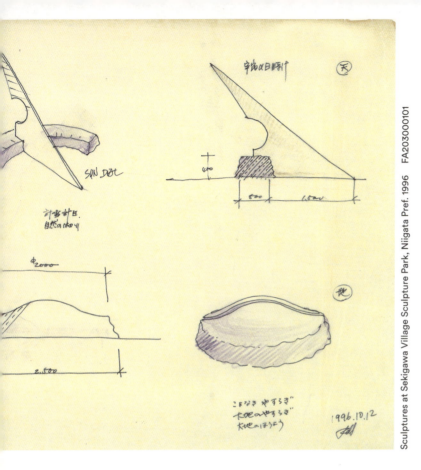

Sculptures at Sekigawa Village Sculpture Park, Niigata Pref. 1996

Drop of Rain rolling down as a 'vortex' shape to the earth.

雨は'うず'となって大地へすいこまれる。

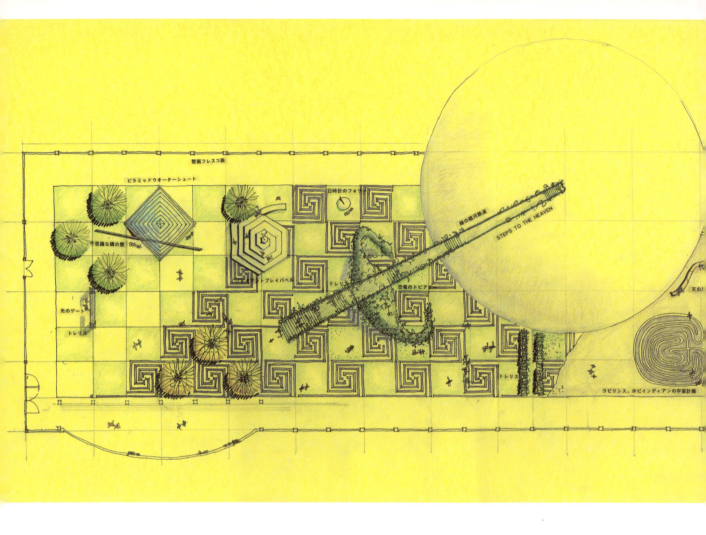

Feel sympathy to the Universe on the Celestial Garden!
The 'vortex' labyrinth of Hopi Indian is map of universe.
Milky way railroad flying to the eternity.

ここは宇宙を感じる屋上庭園。ホピインディアンの'うず'迷路は宇宙図だ！
銀河鉄道はどこまでも続く。

Rest in the Gazebo of Ammonite in the courtyard garden!
'Vortex' of Ammonite is an icon of this land,
used to be the seashore.
アンモナイトのガゼボ（あずまや）でひとやすみ。
'うず'は、むかしここが海辺だったしるし。

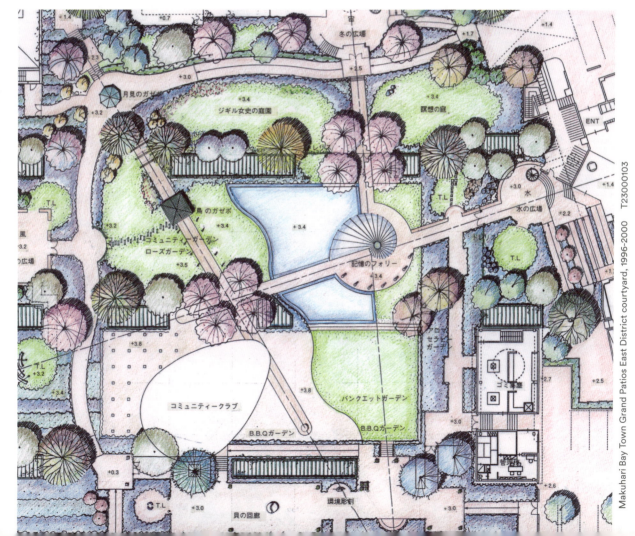

Celestial Garden at Hitachi Science Museum, Hitachi 1987   FA107000102

Makuhari Bay Town Grand Patios East District courtyard, 1996-2000   T23000103

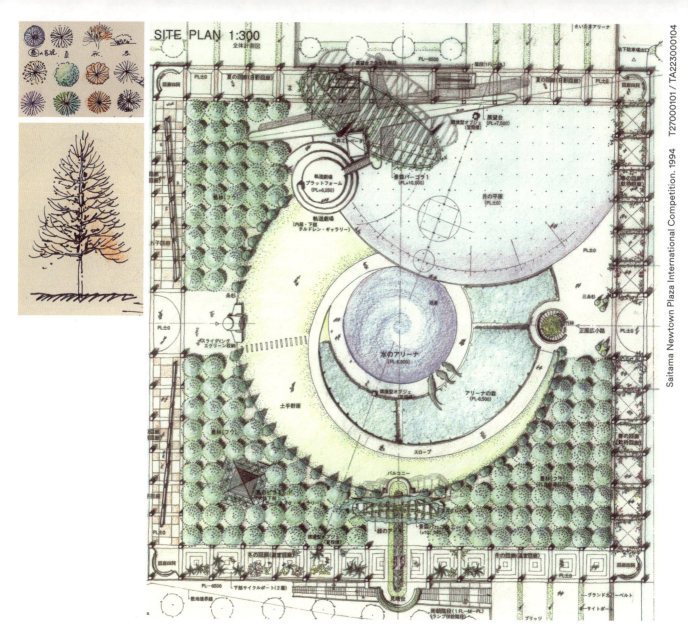

Open air theater of 'Vortex' is located in the center of the forest.

緑のトンネルにかこまれた森。中心には'うず'の野外劇場が!

Here is Nagasaki Port!
Let's land at the big 'Vortex'.
Relax yourself and
look at the sky and the sea!

ここは長崎港！おおきな'うず'に着陸。

ゆったりとねころがって
空と海をながめてみよう！

Mai-Mai Theater at Land Plaza, Nagasaki Seaside Forest Park, 2000-2004 (Nagasaki Urban Design Committee)   T32000101

Yutaka HIKOSAKA
Beyond Vortex

In the form of a whirlpool, everyone can feel the fundamental movement of the underlying vitality and the unique beauty. Vortex narrates time. Time is a surrogate concept of life.
The universe, the atmosphere, the water, and the creatures make vortices. It may be better to say that whirlpools are ubiquitous in the heavens and the earth, or that everything is contained in various eddies.
Not only nebula, typhoon, cloud eddies, tornadoes, eddying tide, sand whirlpools, but streams and the air around us swirls. Snails, cochleas, sensory organs, leaves and petals of many plants, algae also grow along with the vortices, creating vortices.
Aircraft can fly through eddy currents. Vortex sounds music. The vortices diminish even in a coffee cup. The vortices cast deep shadows on mythical labyrinth, ritual ruins, geoglyph, and the Tower of Babel.

Archimedes and Fermat devised their own vortex forms, and Da Vinci designed a centrifugal spiral pump. Paul Valéry has a deep bond with this unique and generating form of the shell, completely natural, which at once could be perceived in a single glance. Both Klimt and Hundertwasser were fascinated by the mysterious vortex. Moreover, James Watson and Francis Crick clarified that the foundation of life is the vortex of DNA double helix.
The Yin and Yang graphism and the shape of the jade represent the affinity between the universe and the vortex. Also, Karman's vortex street a world of diverse spiral figures that is close to the divine wisdom of beyond humans.

A vortex is indisputably an icon of the universe.

A vortex is a spatial form that seals and represents time. It visualized the wise providence of all creation.
The vortices also imply the existence of invisible axiality emanating from its center.
What is beyond there?

Earth, water, and swirls created by the wind become the vortex of space and connect the heavens and the earth. The vortex built into the landscape of Ryoko Ueyama should also be interpreted in that context. At the same time, there may be a glimpse into the creation of Eros covering the environment as a whole, which will transcend rationality into the environment, and the will to put its origin to into the vortex.

彦坂 裕
渦の彼方

渦という形象には、根源的な生命の営みと動き、そして固有の美しさを感じさせる。渦は時間を語りかける。時間とは生命の代理概念だ。
宇宙は、大気は、水は、そして生き物は、渦をつくる。天と地には渦が遍在する、いや、万象は多様な渦に包含されていると言った方が正しいかもしれない。
星雲、台風、雲渦、竜巻、渦潮、砂嘴、渓流のせせらぎも、われわれの回りの空気も渦をまく。巻貝、蝸牛、聴覚器官、多くの植物の葉や花弁、藻も渦に沿って、渦を生み出しながら成長する。
渦流は飛行機を飛ばす。渦は音楽を響かせる。珈琲カップの中にさえ渦は消長する。神話上の迷路や祭祀の遺構、地上絵、バベルの塔にも渦は深い影をおとす。

アルキメデスやフェルマーは独自の渦形態を、ダ・ヴィンチは渦巻き式揚水機を考案した。ポール・ヴァレリーはこの形成的ではなく生成的な貝殻の形姿、人為の入らぬ自然の、しかも一瞥で認識できるその形姿に深い憧憬を抱いた。クリムトもフンデルトヴァッサーも神秘を奏でる渦にとり憑かれた。そしてジェームズ・ワトソンとフランシス・クリックは、生命の基盤がDNA二重螺旋の渦であることを解明した。
陰陽五行のグラフィズムや勾玉の形は、宇宙観と渦の親和性を表現する。そしてカルマンの渦列は、人智を超えた神智にも近い模様の世界を形づくる。

渦は、まごうことなき宇宙のアイコンである。

渦は時間を封印し、かつ時間を表象する空間的形態である。万物創生の摂理が可視化されている。
渦は、その中心から発する不可視の軸性の存在も暗示する。
その彼方には何があるのだろうか?

大地や水、風のつくりだす渦は、宇宙の渦となり天と地をつなぐ。上山良子のランドスケープに内蔵された渦も、その文脈で読まれるべきものだろう。同時にそこには、合理性を超越して環境に瀰漫していくであろうエロスの創造、その起点を渦に託す意志を垣間見ることができるのではないか。

Archives
Cosmophilia
コスモフィリア

Ryoko UEYAMA
In search for a "landscape bridges heaven and earth"

After we have completed our journey around the "Vortex", our hot air balloon "Ubiquitous," has arrived on "Archives Island." This "island" is a collection of landscape designs made by Ryoko Ueyama and her colleagues. It is represented by sketches, drawings, and model and project photographs. We will be visiting each work soon, but we are forming groups based on three perspectives: from the sky, from the surface, and from underground.

First, "perspective from the sky: a new layer of the earth." These projects are based on the view point of birds flying in the sky or aircraft flying at an altitude of 10,000 meters. You can find places where the sea meets earth and where nature and man coexist. Some projects draw a magnificent pattern on earth.
Second, "perspective on the surface: ADORE to the Surface of the Land." These projects emphasize the "water system," such as seas, rivers, and ponds; the "wind system," such as the flow of air and its passage and the undulation of topography; and the "ground system," such as the power of earth, forests, vegetation, and the memory of the people who live there.
Third, "perspective from inside of the earth: imagination and respect to the underground." These projects are based on the recognition that nature is embedded within the earth for hundreds of millions years. They represent respect for its power and threat, the social memory buried underground and the axis from the center of the earth to the sky.
The three perspectives are from those of birds, dogs, and moles. Each work is created from these perspectives, which are closely connected to one another.

"The landscape bridges heaven and earth." This statement suggest that the area of landscape architecture, as if to "weave one cloth" by applying the "weft to the warp" of each specialized field, is required a wide range of knowledge, wisdom, and cosmic sensibility. It is the act of accruing new cultural strata along the time axis to carve some place on earth. For this act, the resonance with the sun, the moon, and all mythical celestial universes, the memory of the land which the human activity has been spun, and the harmony with light, wind, water, soil, flora, and fauna must be very important.

上山良子
「天と地をつなぐランドスケープ」を求めて

「渦」をめぐる旅を終えた今、私たちの熱気球Ubiquitous号は「アーカイヴス島」にやってきた。この「島」には、上山良子とその仲間たちによるランドスケープデザインの仕事が集められ、スケッチ、ドローイング、模型写真とプロジェクトの写真で表現されている。これからその作品をひとつずつ訪れていくことになるが、「空から」、「地表から」、そして「地中から」という3つの「視点」を基軸にグループを作っている。

まず、「天空からの視点：地球への新しいレイヤー」。空を舞う鳥や高度1万メートルを飛ぶ航空機からの視線を、手がかりにしたもの。海と大地が出会う場所、自然と人間が共生する場所、大地に壮大な「文様」を描くプロジェクトである。

次は、「地表からの視点：大地への愛」。海や川や池などの「水の系」、空気の流れや通り抜けの「風の系」、地形の起伏や地質、そして大地と森の力、植生などの「地の系」を踏まえ、そこで暮らす人々の営みの記憶を生かし、地上からの視点を重視したプロジェクトである。

最後は、「地中からの視点：地中への想像と畏怖」。地球という天体の内部に込められた数十億年の時間軸によって自然を捉え、その力の脅威への敬意、地下に埋蔵された社会的記憶、さらに地球の中心から天空へ向かう軸線を体現したプロジェクトである。

この3つの視点は、「鳥」、「犬」、「もぐら」の視点でもあり、どの作品も3つの視点を踏まえて創造され、密接につながりあっているのはいうまでもない。

「天と地をつなぐランドスケープ」。この言葉は、「ランドスケープアーキテクチャー」の領域が、それぞれの専門的な分野の縦糸に横糸をかけて一つの布を織っていくように、広範囲な知識と知恵、ある種宇宙的な感性が要求されることを表している。

地球という大地の一つの場を穿つという行為は、太陽、月、そしてあらゆる神話的な天体宇宙との共振、それまで紡がれてきた人間の営みである土地の記憶の上に立って、光、風、水、土、動植物と共生しつつ、時間軸にそって、新たな文化的な地層を重ねていくことなのである。

From the Sky
A new Layer to the Earth
天空からの視点
地球への新しいレイヤー

Listen to the voice of The Other World

Threshold: the Sea and the Earth
海と大地が出会う閾(しきい)

Y's Ideal Residence at Omura Bay Schematic plan, 1988　T13000203

A Testimony to love nature and to live

自然を愛し生きる証

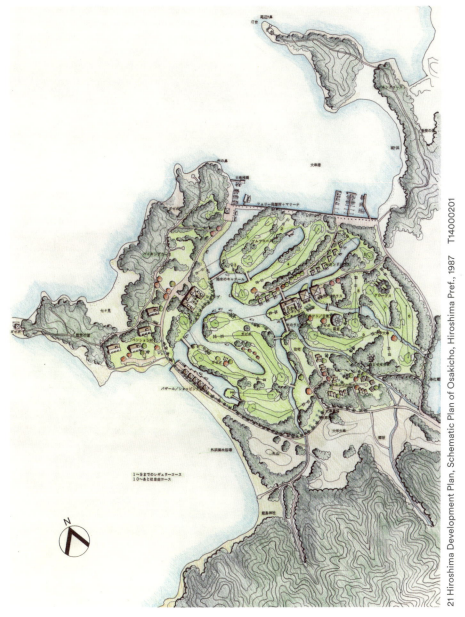

Enjoying the threshold of the Sea and the Land
in playing golf at Setonaikai
瀬戸内海、海と陸との出会う場所でゴルフ

Welcome luxurious liners
on dawn at lawn covered embankment
芝生の土手に迎えられる暁の豪華客船

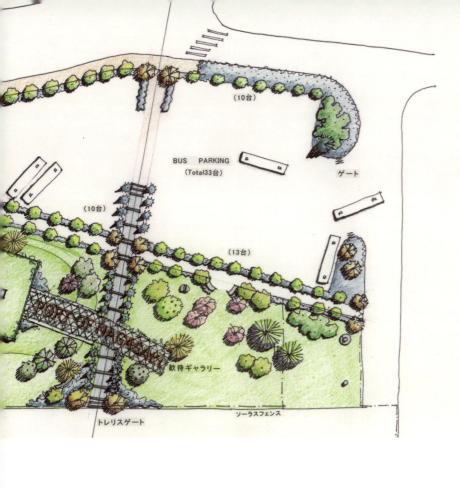

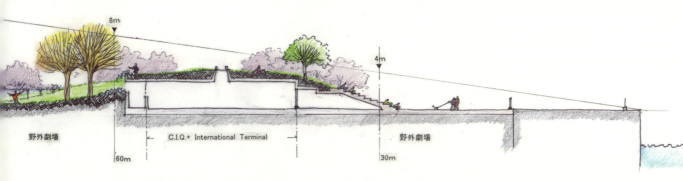

Symbiosis with Nature
自然との共生

Village to learn symbiosis from Agriculture
農に学ぶ共生思想体験邑

New Hiroshima Garden Airport City Schematic Design, 1989　T10000201

LAND ART: The 21st century airport as a Garden City

ランドアートとしての空港庭園都市

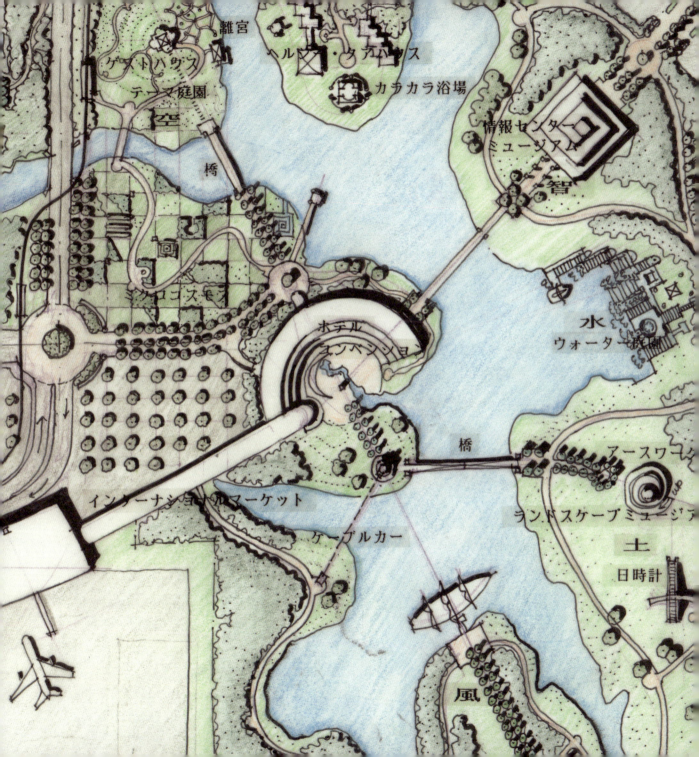

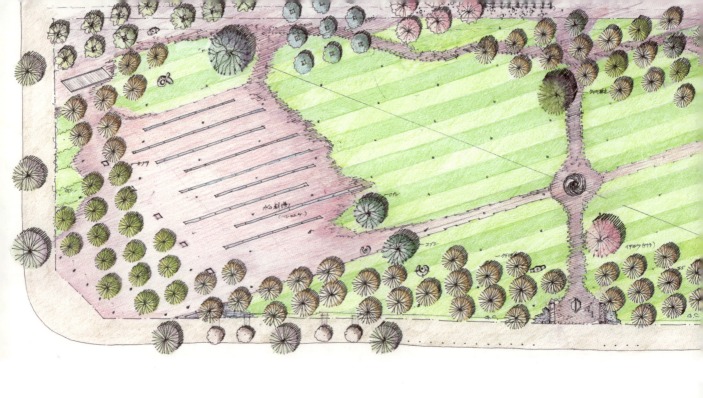

AXIS on the Earth
地球上の軸

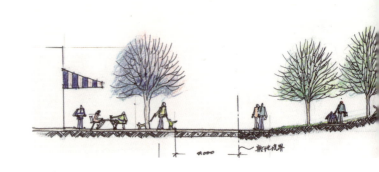

Axis of Mt. Fuji and Mt. Tsukuba
富士山と筑波山への軸線生かす

On the Surface
ADORE to the Surface of the Land
地表からの視点
大地への愛

Listen to the Genius Loci!

Exteriorization of the Water Vein

水脈の顕在化：川を生かす

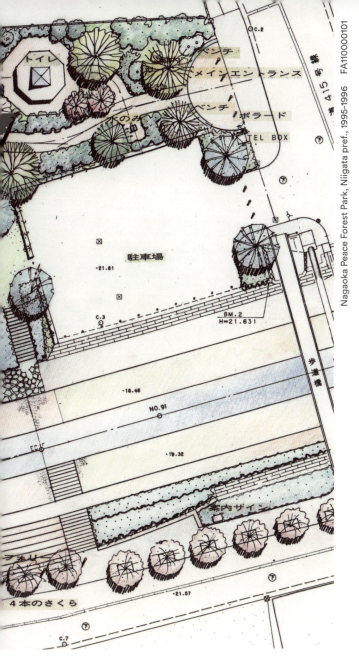

Water Theater, Homage to those who passed in the river in 1945
水の劇場：1945年にこの川で亡くなった人々へのオマージュ

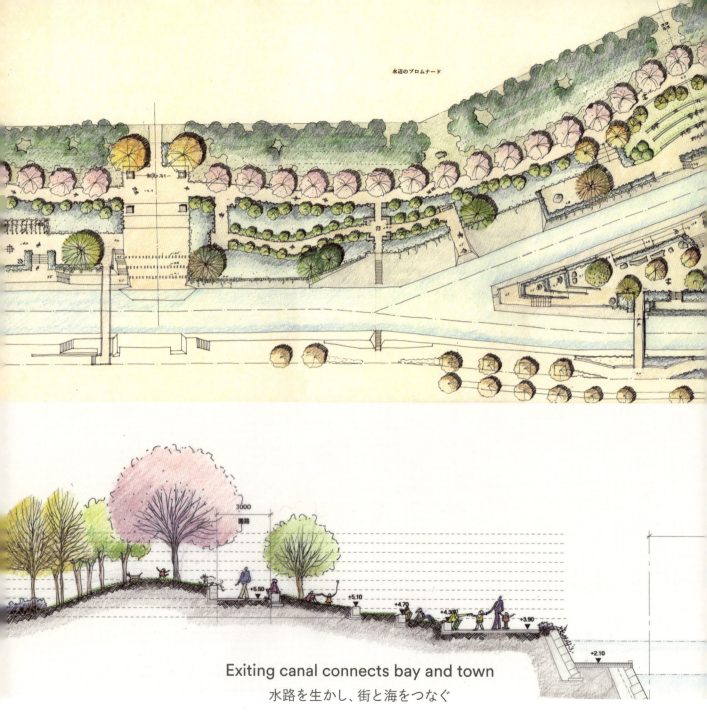

Exiting canal connects bay and town
水路を生かし、街と海をつなぐ

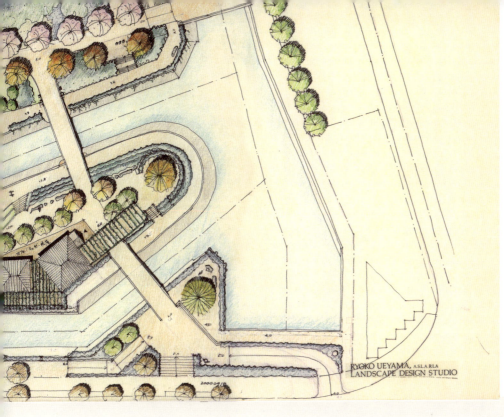

Power of the Forest and the Earth
森の力、大地の力を生かす

Every garden in the Living Units as Green Spine connects to the Park

すべての庭は公園につながる緑の背骨

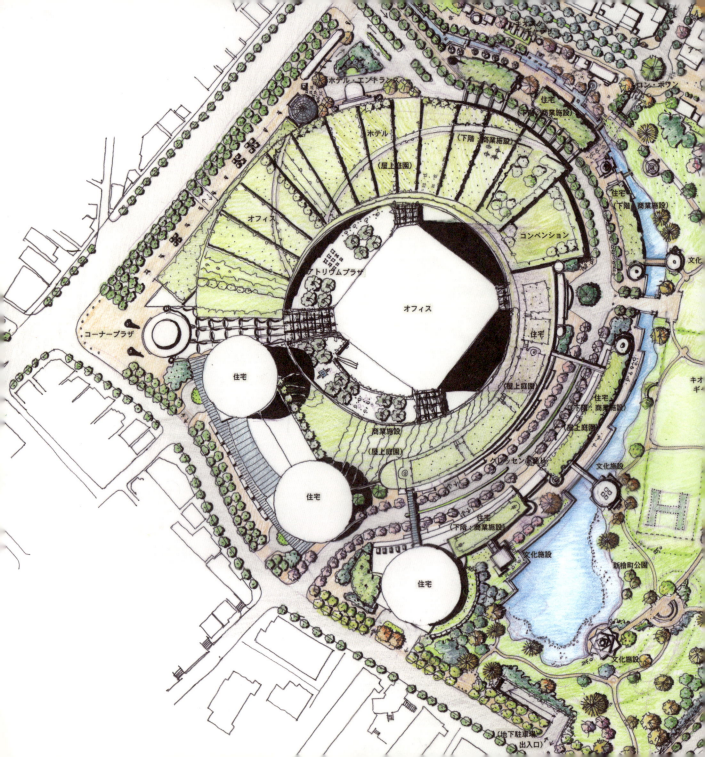

Celestial Garden of Babel vs
the Garden of Memory of the Land
バベルの空中庭園 vs 土地の記憶の庭

Memory of Human Activities
人の営みの記憶

Vocabulary of the British Colony ages ties Time
時をつなぐイギリス植民地時代の語彙

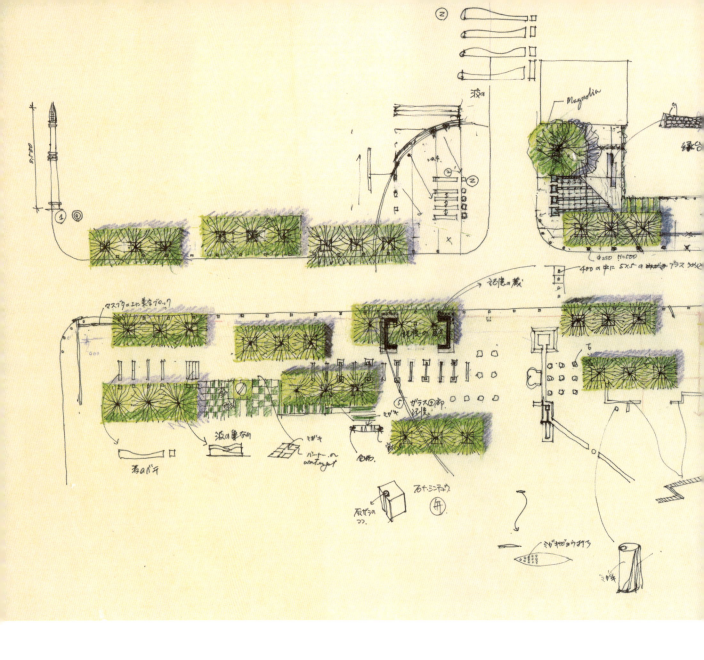

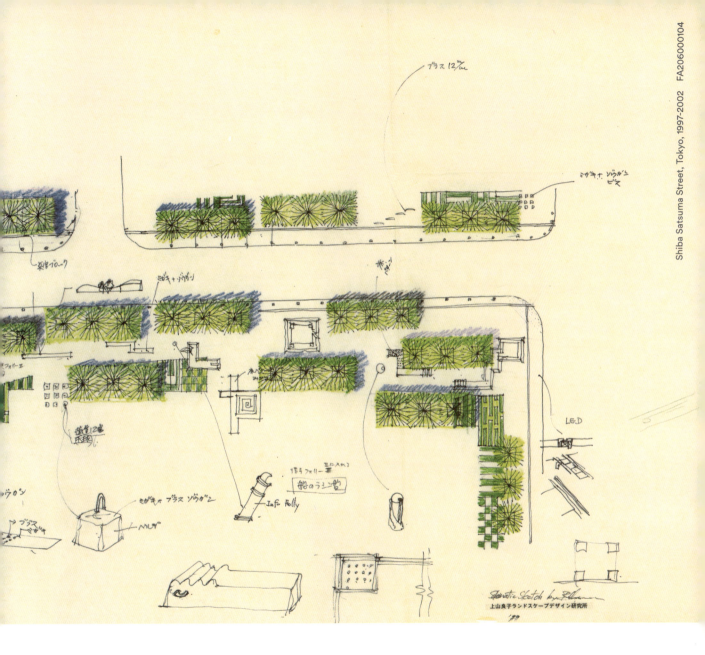

Street is the Open Museum of memories of Satsuma-yashiki
街路は土地の記憶のミュージアム

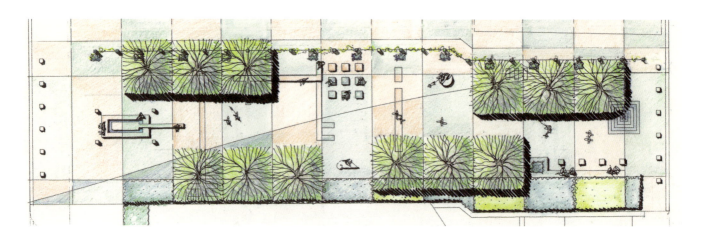

Shiba Satsuma Street, Tokyo, 1997-2002   FA206000104

Street is the Open Museum of memories of Satsuma-yashiki

街路は土地の記憶のミュージアム

To respect Genius Loci of the ancestral Land and Inari shrine
祖先伝来の地霊、稲荷神社を敬う

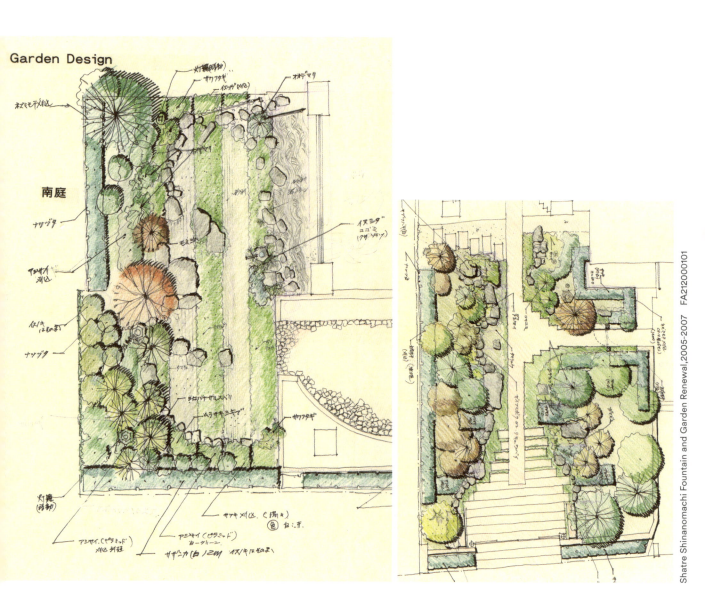

Old Garden survived as a Passage of Time to Future
古い庭を未来に向かう時の通路として再生する

The memory of land invites the innovative science
土地の歴史・記憶が最先端科学を誘う

**From Inside of the Earth
Imagination and Respect to the Underground**

地中からの視点
地下への想像と畏怖

Axis to the Sky from center of the Earth

The Miracle and Menace of Nature
自然の脅威

"Heaven" is the eternal revealed in time: "Earth" is the natural
environment in its broadest ecological sense and
"Wisdom" is the gift that allows humanity to connect "Heaven" and
"Earth" through the artfully designed environment.
Disaster occurs when any of these 3 "Protagonists"
lose sight of each other.
That is when Hazard meets Vulnerability.
The only way to overcome environmental vulnerabilities and
to counteract natural threats is the creation and sustenance of
Landscape as unity of: "Heaven, Earth, and Wisdom".
Insight into the time axis, listening to the voices of the earth,
and having wisdom that transcends personal desires.

「天」は時間、「地」は大地の生態である人間環境、「人」は叡智。
災害は危機と脆弱性が出会うときに起こる。
環境の脆弱性を克服し、自然の脅威と折り合う唯一の方法、それは
時間軸を洞察し、大地の声を傾聴し、私欲を超越した叡智を持つこと。
それが「天・地・人」が合一したランドスケープ。

Metaphor of the Memory of Land lying in Underground
地下に眠る土地の記憶を暗示する

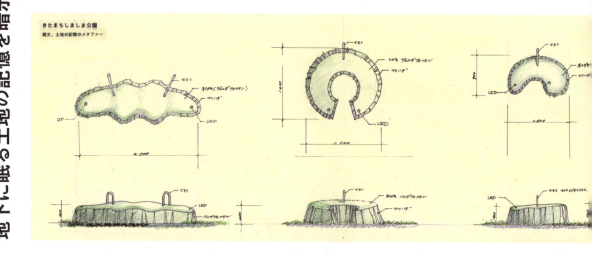

Kitamachi Shima Shima Park, Saitama Pref.,2001-2005 FA121000102 / D070007s

The Sculptures of the seat as the metaphor of the memory of Jomon Era scattered ubiquitously on a lawn

縄文時代の遺跡を暗喩する座の彫刻を偏在させ、祖先の声を聴く

Axes from the Centrosphere to the Heavens
地球の中心から天空への軸線

Nagaoka Peace Forest Park, 1996

Create monumentality as a space for craving Peace
平和を希求するモニュメンタルな空間を創る

Nagasaki Seaside Forest Park, 2005   D070006s

Axes to the memorial places of Nagasaki and the Bridge are crossing in the park
The symbolic vortex in the center spreads to the Universe

長崎の土地の記憶と海につながる橋への軸線が公園で交わり、世界に広がる渦になる

Axes to Mt.Fuji and Mt.Tsukuba are visualized to inherit the Memory of the Land to the Future

富士山と筑波山へ向かう軸線を顕在化し、大地の記憶を未来につなぐ

Roof Garden in Hitachi Civic Center Science Museum, 1987   D070001s

Celestial Garden connects Heaven and Earth
天と地をつなぐ空中庭園は、時空をこえた宇宙との交感の場

Makuhari New Town Gran Patios Park East Town, 2000 D070008s

Dramatic Stage Set in Urban Resort Livelihood
リゾート感覚の舞台装置が都市生活の風景に潤いをもたらす

Shiba Satsuma Street, 2002   D070005s

Street is the "Museum" to exhibit 'The Memory of the Land'
「土地の記憶」を生かした道そのものがミュージアム

Ryoko Ueyama Exhibition "Vortex"

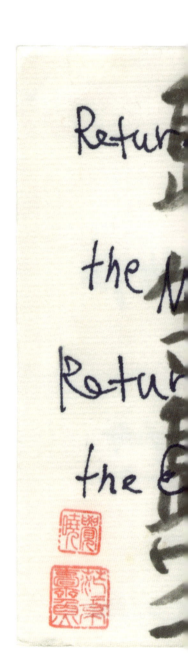

What I feel from the works of Ryoko Ueyama is the wonder of the form of "Water". Even if the real water is not used at all, I am attracted by stone formations, soil undulations, shapes of garden path, and the measured arrangement of paving stones that remind me of the water flow, stagnation, and "vortex".
As a famous sentence in Hōjōki of Kamo no Chōmei, "The flow of the river which goes continuously and is not the original water", the water flow is plainly embodied "everything is rolling", and "eternal return".
For Ryoko Ueyama "the concept of creating a place" is "cosmophilia", which, I think, is the recognition of the universal truth of "all things must pass" and "eternal return". She also says that she was enlightened by Yi-Fu Tuan's concept of "topophilia" when she first pursued landscape design.
For her, "Topophilia" is the inevitable result of explorations within an underlying concept: *creating a landscape as a 'place' for human beings,* from the perspective of macroscopic "cosmophilia". Herein, the expression of "water", especially "vortex" derives from this process.

Here's superfluous but, recently, I needed to re-read Descartes's *"Principia Philosophiae* : Philosophical Principle", in which theory of the "vortex cosmology" is described. It is a theory that all the movement in the universe has been explained by the vortex movement of the particles that are filled with the universe. And it is pioneered the cosmology of quantum "field" of the present days. However, I thought it was interesting that there was something that associate with "vortex" of Ueyama.

Kakugyo S. CHIKU
Specified Non-profit Corporation for the Inheritance of Architectural Culture
KIT Research Institute for Architectural Archives

上山良子「渦」展に寄せて

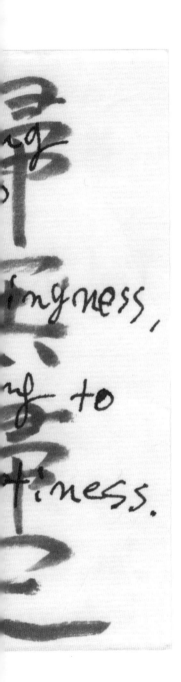

上山良子さんの作品を観ていて、私が感じたのは「水」の造形の素晴らしさである。水が全く使われていない場合でも石の造形や土の起伏、園路の形状や石畳の配置などに、水の流れ、澱み、そして「渦」を彷彿させるものがあり、それが私を強く惹きつけた。鴨長明の方丈記の「ゆく川の流れは絶えずして、しかももとの水にあらず」という有名な文章があるが、水の流れは端的に「万物流転」「永劫回帰(eternal return)」ということを体現している。上山さんは「場づくりのコンセプト」は「コスモフィリア(cosmophilia)」であると仰っておられるが、これは「万物流転」「永劫回帰」という宇宙的真理の認識であったと思う。上山さんはまた、自身がランドスケープデザインを始めた頃に、イーフー・トゥアン(Yi-Fu Tuan)の「トポフィリア(topophilia)」なる概念に啓発されたと仰っておられる。私は上山さんが、巨視的な「コスモフィリア」を微視的なランドスケープ、すなわち、人間の「場」として造形するための契機が「トポフィリア」だったのであり、それが「水」「渦(vortex)」という表現に帰着したのではないのか、と思っている。

以下は蛇足だが、最近、必要があってデカルトの『哲学原理』を読み直したのだが、そこで展開されている「渦動宇宙論」は、宇宙に満ちている微粒子の渦動によって宇宙内の運動一切を説明したものであって、現在の量子論的「場」の宇宙論を先駆した論である。これはしかし、上山さんの「渦」に通じるものがあると思って面白く思った。

竺 覚暁
NPO法人 建築文化継承機構
金沢工業大学建築アーカイヴス研究所

Epilogue

Water: Ultimate Beauty
水：究極の美

main basin

the sharper the edge.
linear water effect.

With your Six senses!

FA223000104

A Scene at the contact point: Water and Other Materials
水と他の物質との接点に生じる景

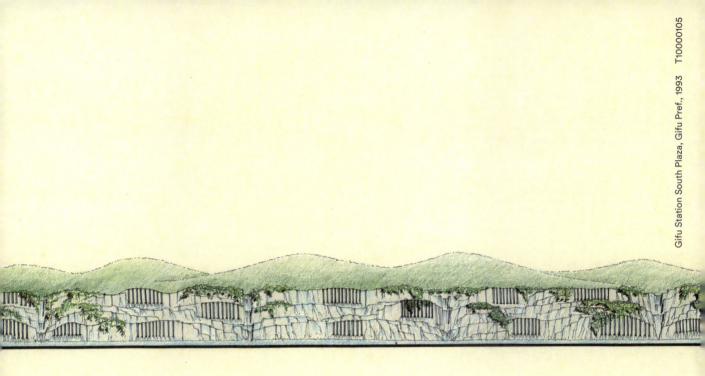

Gifu Station South Plaza, Gifu Pref., 1993 T10000105

**Infinite variety of Water trickling down from Rocks
Symbol of the Scenic Beauty of Gifu**

岩から滴る水:千差万別の趣。山紫水明の岐阜の象徴

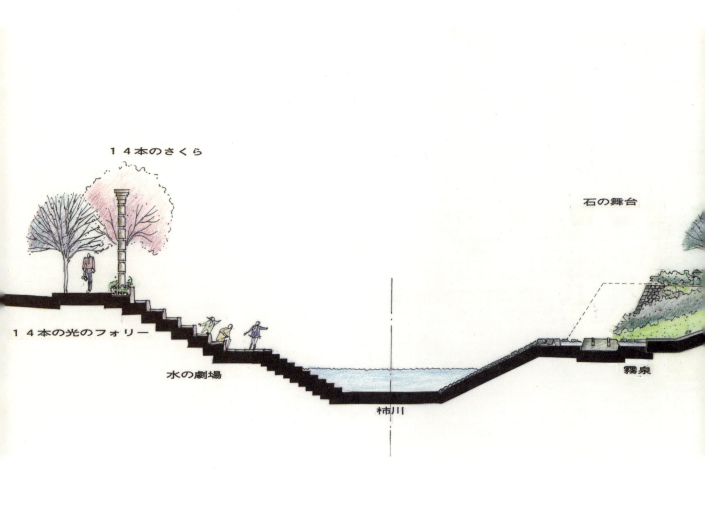

Fountain of Water Basin: Tsukubai reflects the Universe,
the water is flowing down to the River of memory
「つくばいの泉水」は天を写し、流れ落ちる水は記憶の川へ

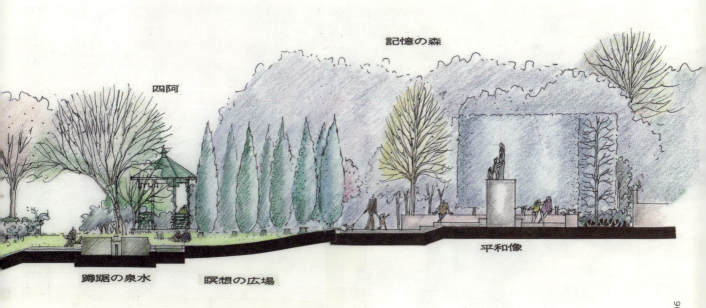

記憶の森
四阿
平和像
蹲踞の泉水
瞑想の広場

Tsukubai-Fountain, Nagaoka Peace Forest Park, 1996

Water as metaphor of old wells
古井戸のメタファーとして

Three Fountains at Shiba Satsuma Street, Tokyo, 2002

"Well of the Memory"
「記憶の井戸」

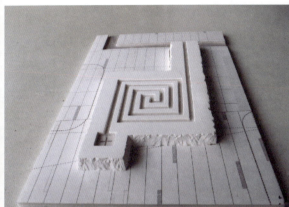

"Well of the Labyrinth"

「迷宮の井戸」

Water as a Mirror vs Falling Water
鏡としての水 vs 落水

Infinity: Continuation of the Surface of Water to the Sea
インフィニティ：海へつながる水面の連続

Place of Learning from Water
Children can aware of the wisdom of the nature with water
水に学ぶ場所：こどもたちは水に触れて自然の叡智に目覚める

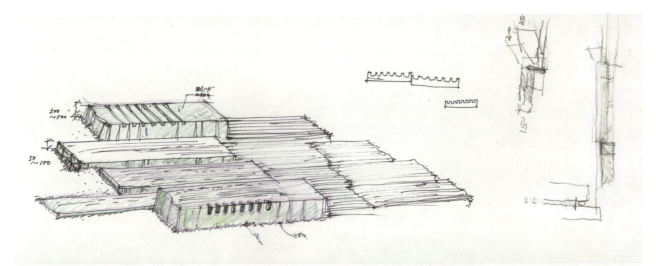

Cycle of Nature: Water returns to the Sea and ascends to Heaven
自然の循環：水は海に帰り天に昇る

Kazuaki SEKI
Memory of the place and History of Universe bridge to "the other world"

What a drawing tells

First, let us appreciate the beauty of Ryoko Ueyama's hand-drawn illustrations.
Ms. Ueyama's landscape design are drawn, colored, and shaded by her own hand on large tracing papers. These precise drawings and sketches represent everything from the original concept to the form of details. All the drawings are donated to the Research Institute for Architectural Archives, Kanazawa Institute of Technology and preserved by experts in this institution.
Landscape design drawings, in which most of the subjects are outdoor (i.e., a place not covered by a roof), differ from architectural drawings. The buildings, trees, undulations in the land, people, and vehicles are drawn from the perspective of "looking down from above." The sections, elevations, and perspective drawings are more similar to pictures than to drafted figures with numerous abstract signs.
Apart from its point-of-view and method, Ms. Ueyama's work has another important feature. That is, every drawing narrates "a story of the imagination that spread to the universe."

"Myth" of nature and human beings

What is this "story" like? It is not a story that people in modern times immediately think of; instead, it is a timeless, archaic, and primordial story that may be called "myth."
From ancient times until the present, the metamorphoses of the classical elements that make up the world (cosmos), namely, water, soil (earth), wind (air), and light (fire), have inspired various stories. In her projects, Ms. Ueyama does not only use abstract materials but also natural objects (e.g., celestial bodies), topography, the environment (e.g., plants and animals), and all human activities (e.g., society and urban space) as characters to play roles in the important moments of the "story". It is strongly anchored to the human body and the senses.
Ms. Ueyama had experienced looking at the ground every day from an altitude of 10,000 meters in different parts of the world before she became a landscape architect. The sensibility that is inherent in the view from the stratosphere complements the view on the ground surface and eventually provides the power to create unique ideas and designs.
Dr. Yi-Fu Tuan coined the word "topophilia" to refer to "love for a place" (i.e., emotional engagement to a specific place). By contrast, Ms. Ueyama advocates the concept of "cosmophilia," i.e., love for universe, which extends the concept of topophilia to seek the sublime. Interposing imaginative distance (space and time) can capture the diverse aspects of the Earth. The vision of the Earth beyond the endless heaven creates a landscape that only exists there, with an intimate observation of events (i.e., history and life) on the ground.

Reincarnating the memory of the land

At the center of "Nagasaki Seaside Forest Park" adjacent to the port, we can see a huge vortex drawn on the ground. This figure, which is both centripetal and centrifugal, metaphorically represents the dual cultural characteristics of this port city, namely, the acceptance of foreign culture

関 和明
場所の記憶と宇宙史を 'the other world' に架橋する

ドローイングが物語るもの

まず、上山良子さんの手描きドローイングの圧倒的な魅力を存分に感じよう。大きなトレーシングペーパーに、フリーハンドで緻密に描きこまれ、彩色や陰影の施された図やスケッチには、上山さんのランドスケープデザインのオリジナルなアイデアから細部の造形にいたるすべてが表現されている。これらは、JIA-KIT建築アーカイヴス委員会に寄贈され、そこに収蔵されている。

ランドスケープデザインのドローイングは、建築の図面のように記号化された図形の集積ではなく、かつ、描かれているのは、ほとんどが、「屋外」(屋根で覆われていない場所)である。建物も樹木も土地の起伏も人や車も、真上から見下ろした姿で描かれ、断面図や立面図、透視図なども図面というよりは絵画に近い。

しかし、上山さんのドローイングには、単に視点や描写方法だけにとどまらない大きな特徴がある。それを一言で表現すると、どのドローイングにも「宇宙大に広がる想像力の物語」が常に綴られていることだ。

宇宙大の自然と人間とが綴る「神話」

では、その「物語」とはどのようなものか。それは、近現代人がすぐ思い浮かべるような物語ではなく、時間を超越したアルカイックなもの、原初的な「神話」といってよい。太古から現在まで、水や土や風(空気)、そして光(火)など、世界(コスモス)を充たすエレメントの変容が多彩な物語を綴ってきたが、彼女のプロジェクトが綴る「物語」では、抽象的なマテリアルだけでなく、自然界の天体、地形、植物や動物などの生態、そして人間界の社会や都市空間などの人間の営みも、「物語」を紡ぎ出す重要な契機であり、現実の身体や感覚と強くアンカーされている。

「高度1万メートル」から地上をみること。上山さんは、ある時期、世界中の場所で日常的にこの体験をしていた。ランドスケープ アーキテクトを志す前のことだが、成層圏からの視野を内在した感性は、地表からの視線を補い、やがて独自のデザインについての考えと造形を創り出す力になった。

イーフー・トゥアンは「場所への愛(情緒的な関わり)」を指して「トポフィリア」という言葉を造ったが、上山さんは「トポフィリア」を包括しつつそれを拡張して、崇高さをも希求する「コスモフィリア:宇宙・愛」を提唱する。地上の様相を捉えるために、壮大な想像的な距離(空間の、そして時間の)を介在させること。果てしない天空の彼方から地上に注がれる視線は、地上の場所での出来事(歴史や生活)への親密な眼差しと交錯しつつ、「そこ」にしかない風景を創造する。

土地の記憶を蘇生させる

港に接する「長崎水辺の森公園」の中心には、遠くから見える巨大な渦巻きが描かれている。求心的でもあり遠心的でもあるこの図形は、港町:長崎の文化の両義的な性格(外来文化の受容と固有文化の発信)を暗喩的に物語る。

かつて武家屋敷があった都心の高層ビルの間に作られた街路「芝さつまの道」では、古絵図に記された井

and the dissemination of indigenous culture.

In "Shiba Satsuma Street," which is located between skyscrapers in a city, a samurai house with wells was once depicted in an old map as a design motif. Mysterious and familiar water scenes appear in the public space of the metropolis.

A large double spiral that purifies the water in a lake is proposed to the international competition "Toolonlahti Lake Landscape Design" in Helsinki. This major project dynamically reorganizes the lake's surface changing in summer/winter, the surrounding environment, urban functions (e.g., cultural and living facilities), and the traffic axis of the city. Various "parkette: a park in the park" areas are scattered around the lake.

Apart from the aforementioned representative work, the other drawings of Ms. Ueyama have the following elements in common: vortices, water, stones, and large figures drawn on the ground. These elements symbolize the memory of a place. They are metaphors and creations of the "new layers" as carefully layered monumental objects. These "layers" do not hide currently existing objects but reflect the "memory of the land" that we have never been reminded previously.

"Ut pictura poesis: poetry like a picture." This phrase suggests the idea of poetry (literal expression) handed down from ancient times. "Poetry is like a picture, and a picture is like a poem." Paradoxically, however, we have moved toward the direction of cutting off the link between the visual and the literal in this modern age. Nevertheless, I believe that Ms. Ueyama's work on landscape architecture leads to questions regarding this process because for her, the primary issue is values, which should be considered before design methods and techniques; that is, the meaning of the spiritual world and the view of the world. This manner of thinking distinctly differs from the values of the modern age and/or the contemporary society. However, it does not lead toward the so-called "post modernism," which became popular in the 1970s–1980s. Ms. Ueyama strongly believes that landscape architecture is the art created through imagination. It can reach out to generations and traverse from the beginning of the Earth in the primordial times to the metamorphoses of nature (memory of the universe) and the history of humanity (memory of a place) that span time and space over billions of years.

"The return of the beautiful female gardener"

In the places designed by Ms. Ueyama, various objects are disseminated, giving the impression that numerous fragments are randomly scattered. However, the users of this place are entrusted to find the threads that connect these fragments.

The type of "story" wove by seeing, listening, and reading should depend on the freedom of the one gaining the experience, which should not be limited to adults but should also include children and other creatures (e.g., birds, dogs, and moles). Consequently, countless "stories" are generated. Through the symbols of vortex and water, people are unconditionally free to enjoy these "stories" and to recognize whether they represent the "supreme existence" or are metaphors of "Eros" (derived from feminine).

In any case, through land art depicted on the canvas of the Earth, our souls are bridged to "the other world" (spiritual values).

戸をモチーフにして、不思議でかつ親しみやすいさまざまな「水景」を、大都市の公共空間に現出させた。ヘルシンキのトゥーロンラッティ湖周辺を再生する壮大なプロジェクトのコンペ案（共同設計）では、湖水の水質浄化を兼ねた二重の大きな渦巻きを中心に造り、夏／冬で変貌する湖水面と、それを取り囲む自然、文化・生活施設などの都市機能、そして交通軸をダイナミックに再編成して、湖の周囲の場所に多様な「公園の中の公園」（「パルケット」）の領域を点在させている。

これらの代表作だけでなく、上山さんの他の作品に共通するのは、渦や水、石や地表に描かれる大きな図などのすべてが、その場所の記憶を象徴化し、暗喩となり、モニュメンタルなオブジェクトとして「新しいレイヤー」を創り出し、それを丁寧に重ねていることだ。その「レイヤー」は、いままであるものを覆い隠すのではなく、いままで感じられなかった「土地の記憶」を顕現させるのである。

「ut pictura poesis：ウト・ピクトゥラ・ポエシス：詩は絵のごとく」。古代から伝承された、「詩」（文章表現）の理想を語った言葉である。後代、「詩は絵のように、絵は詩のように」と敷衍されたが、逆説的なことに、近代は視覚的なものと言葉との連関をあえて切断し、抽象化する方向に進んだ。しかし、上山さんのデザイン（ランドスケープアーキテクチャー）は、この「切断」に対して大きな疑問を投げかけているのではないか。というのは、デザインの手法や技法の手前に横たわる「価値」、精神世界の意味や世界観の存在を、彼女は第一義的な課題としているからだ。これは、近代、あるいは現代社会を主導する価値とははっきりと異なる。といっても、1970-80年代に流行したいわゆる「ポスト・モダニズム」と同調するようなものでは全くない。ランドスケープアーキテクチャーとは、原初における大地の生成と流転、数十億年を超える時間と空間のスパンで生じる自然の変容（宇宙の記憶）と人為の歴史（場所の記憶）に想像を届かせることによって、はじめて成立する「アート」であると、彼女は確信しているからである。

「美しき女庭師の帰還」

上山さんによってデザインされた場所には、さまざまなオブジェが点在していて、一見すると多数の断片がランダムに散種された印象を受けるかもしれない。しかし、それらを結ぶ糸を発見しつなげることは、この場所の享受者にゆだねられている。綴られた「物語」を「見ること」、「聴くこと」、そして「読むこと」によって、どのような「物語」を紡ぐかは、体験する者の自由である。大人だけでなくこどもや人間以外の生物（鳥、犬、そしてもぐらたち！）によって、複数の、あるいは無数の「物語」が生成される。「渦」や「水」から「至高存在」の表象を認識するのも、「（女性性に由来する）エロス」のメタファーを感受するのも、享受者の自由だ。だが、いずれにせよ「大地という（空間的な）キャンバス」に描かれたランドアートによって、私たちの魂は、"The other world（精神世界）"へ架橋されるのである。

Acknowledgment

Many people have influenced my life directly and indirectly, and I am full of gratitude for what has been given to me in the last chapter of my life.

As I reorganize my works donated to the Institute for Architectural Archives, KIT for the exhibition, I will go on a journey and explore the roots of my thoughts.

Half a century ago, my diploma in the art of tea from Hazama Gyokufu was entrusted to Mr. Kenzaburo Matsumura, a painter and a calligrapher. At that time, working 10,000 meters above the sky was a daily routine of me; it was the place where I worked. Mr. Matsumura had once told to me to "Be aware the origin of your ideas, which is between the sky and the earth." Now I remember what was said and feel my destiny.

As my life took its twists and turns, I gained exposure to landscape architecture and realized the essence of it from Lawrence Halprin, who at the time, was the top American landscape architect. He taught me to find for myself what it was to create a place for people and stand in harmony with nature. Richard Bender, the dean of UC Berkeley's graduate school of environmental studies, whose personal reach and abilities have provided students with special opportunities for learning, was strong influence in my life. Among the noteworthy classes I attended was the environmental sculpture class by Tino Nivola, who was one of the friends of Le Corbusier. The living lessons that I learned from this little but big-hearted Italian had a considerable influence on my creation of places for the people.

I cannot help but thank Prof. Shigeru Ito and many other people for giving a totally unknown individual many opportunities to create after returning to Japan.

I am deeply grateful to each and everyone involved in the "Uzu Nakama: Colleague of Vortex", from long-time companions to young graduates of Nagaoka Institute of Design, who gathered to prepare the exhibition and publication of the catalog. I wish to thank Mr. Kazuaki Seki for providing me with special guidance on editing the catalog. Finally, I would like to express my gratitude to Mr. Kobayashi and Mrs. Yamada of SPREAD who directed both the exhibition and the catalog. I am able to bring them to the world with the help of many people.

I hope that this exhibition and the catalog will reach the eyes of as many children as possible, who will develop an interest in learning, "What is landscape?"

Ryoko UEYAMA

謝辞

生きていくということはどれだけの人々から直接、間接に影響を受け、また、お世話になることかと人生の最終章の今、感謝の念でいっぱいです。この度JIA-KIT建築アーカイヴス委員会へ寄贈した作品を展覧会のために再整理するにあたり、自分の思想のルーツを探る旅に出ることになりました。

半世紀前、茶道の師であった間玉風は、奥伝の免状を、画家で書道家の松村健三郎氏に託されました。当時、高度1万メートルの世界が日常である仕事をしていた私に、松村氏は「自分の発想の原点は天空と地球との間にあることを自覚せよ」と言われたことを思い出し、運命を感じています。

紆余曲折の末にランドスケープアーキテクチャーという分野に出会い、その真髄を自覚させられた人物が当時の米国のランドスケープ業界の頂点にいた、Lawrence Halprin。自然との共生の上に立つ、「人のための場づくりとは何か」を自分自身でみつけることを教えられました。UCバークレー大学院環境学部の当時の学部長であったRichard Benderの人脈の広さと実行力は、この特別の機会を学生たちにあたえたのでした。もう一つ圧巻の特別カリキュラムが、N.Y.で活躍していたTino Nivola（彫刻家）の環境彫刻クラス。ル・コルビュジェの友人であった、この小さくても大きな心のイタリア人、Tinoから受けた生き生きした教えはその後の私の場づくり、人づくりに大きな影響を与えることになりました。

帰国後、全く無名の私を、信じて創造する機会を与えてくださった伊藤滋先生をはじめ、多くの皆様には感謝してもしきれない思いです。

そして、今回この展覧会と図録出版に際し、長いお付き合いの仲間から、今活躍の最中の長岡造形大学の若い卒業生たちにいたるまで、集まってくれた「渦なかま」一人一人に心から感謝の意を表します。図録出版に際しては、格別の指導をたまわった関和明氏に心より感謝する次第です。最後に、展覧会・図録ともにディレクションを担ってくれたSPREADの小林さん、山田さんに感謝の意を表します。

この展覧会と図録に関してはその他多くの方々のお力によって世の中に出すことができました。この展覧会と図録が少しでも多くのこどもたちの目にふれ、手にわたり、「ランドスケープって何？」と興味を持ってくれることを願ってやみません。

上山良子

Ryoko UEYAMA Profile

Landscape architect Ryoko Ueyama returned from the United States in 1982 and founded Ryoko Ueyama Landscape Design Studio in Tokyo. Following her own philosophy, she has been engaged in landscape design for more than 30 years. In a sense, she is a pioneer in this field in Japan.

She graduated with a Bachelor of Arts degree from Sophia University in 1962, and she left to the United States after gaining experience in various design fields.

In 1978, she completed the Graduate Program in Landscape Architecture at the University of California, Berkeley and received the American Society of Landscape Architects Merit Award in the same year.

Upon graduation, she worked as a project designer for C.H.N.M.B. Associates (Former Lawrence Halprin & Associates) in San Francisco until 1982.

While working as a practitioner, she was a professor in the Department of Environmental Design, Nagaoka Institute of Design (NID) from 1995 to 2005. From 2008 to 2012, she was the president of NID. She is currently a professor emeritus of the institute.

As a design practitioner, she has engaged in an extensive range of landscape design contexts and scales. With her fascination with intimate and meticulous detail and her expansive and cosmic inspiration for rural and urban settings, she has received numerous awards and much recognition for her achievements.

Her concepts are to design a space with deep appreciation of "The Memory of the Land" and create the space as "only one place" in the world. In such a manner, she thinks a new landscape of Japan should be created. She is now contributing to spark interest in the field among young children, hoping that they will pursue this meaningful endeavor in the future.

Awards
1996:   "Nagaoka Peace Forest Park" AACA (Japan Association of Artists, Craftsman and Architects) Award.
                2014: Nagaoka Cityscape Award
1997:   "Helsinki, Lake Toolonrahti Park International Competition (Invited)" Perches Award
2002:   "Shiba Satsuma-no-Michi": Good Design Award
2004:   "Nagasaki Seaside Forest Park" Good Design Grand Prix Award
                (Nagasaki Urban Design Committee: Chairman Shigeru Ito).
                2006: Japan Society of Civil Engineers Design Award. 2013: Nagasaki Cityscape Award
2006:   "Kitamachi shima shima Park" Good Design Award, AACA Award
2011:   "Nagasaki Port Matsugae International Tourism Vessel Wharf" Good Design Award
                (Nagasaki Urban Design Committee). 2011: Nagasaki Cityscape Award. 2013: Japan
                Society of Civil Engineers Design Award
2019:   "Nagasaki Urban Design Committee" won the Minister of Land, Infrastructure,
                Transport and Tourism Award at the 100th Anniversary Ceremony for the City
                Planning Act and the Urban Building Law

Publications
LANDSCAPE DESIGN "Creating a Place" - Listen to the Voice of the Earth, Art Publishing Company, 2007.
https://ebookjapan.yahoo.co.jp/books/327421/A001560706/

The INSPIRED LANDSCAPE: Twenty-one leading landscape architects explore the creative process by Susan Cohen, TIMBER PRESS 2015.

上山良子 略歴

ランドスケープアーキテクト上山良子は、1982年に米国から帰国し、上山良子ランドスケープデザイン研究所を設立。30年以上にわたり、独自の哲学を持って、ランドスケープデザインに携わってきた日本におけるランドスケープデザインの草分けと言える存在である。

上智大学英語学科を卒業後、様々なデザイン分野を経験したのち、渡米。1978年カリフォルニア大学環境デザイン学部ランドスケープアーキテクチャー学科の大学院を修了。ASLA（全米ランドスケープアーキテクト協会）賞（大学院の部）を受賞。その後CHNMB事務所（旧ローレンス・ハルプリン事務所）にプロジェクト・デザイナーとして1982年まで参画し、数々のプロジェクトを経験して帰国。

実践家として現場までこなす一方、長岡造形大学大学院造形学部環境デザイン学科にて、1995年から2005年の間は教授として、2008から2012年までは学長として、学生たちを育ててきた。現在は同大学名誉教授。

デザインへのこだわりが来訪者の共感を呼び、数々の賞へと繋がる。土地の記憶を大切にして、場所性を担保し、日本の新たな風土を創ることを目指して、「そこしかない」空間デザインを追求する。現在は日本の子供たちにこの世界の面白さを伝えたいと心から願って、行動している。

受賞歴

1996年	「長岡平和の森公園」 AACA賞受賞、2014年、長岡市都市景観賞受賞
1997年	ヘルシンキ、トゥーロンラッティ湖公園国際コンペ（招待）パーチェス賞受賞
2002年	「芝さつまの道」グッドデザイン賞受賞
2004年	「長崎水辺の森公園」グッドデザイン賞金賞受賞（環長崎港アーバンデザイン会議：伊藤 滋座長）、2006年 土木学会デザイン賞受賞、2013年 長崎市都市景観賞受賞
2006年	「きたまちしましま公園」グッドデザイン賞受賞、AACA賞入賞
2011年	「長崎港松が枝国際観光船埠頭」グッドデザイン賞受賞（環長崎港アーバンデザイン会議）、長崎市都市景観賞受賞、2013年 土木学会デザイン賞受賞
2019年	「環長崎港アーバンデザイン会議」は都市計画法・建築基準法制定100周年記念式典において国土交通大臣表彰を受賞

著書

LANDSCAPE DESIGN 「場を創る」―大地の声に耳を傾ける、美術出版社、2007年。
https://ebookjapan.yahoo.co.jp/books/327421/A001560706/

The INSPIRED LANDSCAPE: Twenty-one leading landscape architects explore the creative process by Susan Cohen, TIMBER PRESS 2015

Contents   目次

02   Ryoko UEYAMA
      Between the Sky and the Earth:
      The last flight of the Hot-Air Balloon 'Ubiquitous'
      上山良子
      空と地球の間にて：熱気球Ubiquitous号の最終フライト

04   **Prologue**
      Vortex: Ultimate Existence
      プロローグ
      渦：究極の存在

22   Yutaka HIKOSAKA
      Beyond Vortex
      彦坂 裕
      渦の彼方

24   **Archives: Cosmophilia**   アーカイヴス：コスモフィリア

26   Ryoko UEYAMA
      In search for a "landscape bridges heaven and earth"
      上山良子
      「天と地をつなぐランドスケープ」を求めて

28   From the Sky: A new Layer to the Earth   天空からの視点：地球への新しいレイヤー

40   On the Surface: ADORE to the Surface of the Land   地表からの視点：大地への愛

58   From Inside of the Earth:
      Imagination and Respect to the Underground
      地中からの視点：地下への想像と畏怖

70   Kakugyo S. CHIKU
      Ryoko Ueyama Exhibition "Vortex"
      竺 覚暁
      上山良子「渦」展に寄せて

72   **Epilogue**
      Water: Ultimate Beauty
      エピローグ
      水：究極の美

86   Kazuaki SEKI
      Memory of the place and History of Universe bridge to "the other world"
      関 和明
      場所の記憶と宇宙史を'the other world'に架橋する

90   Acknowledgement   謝辞

92   Ryoko UEYAMA Profile   上山良子 略歴

Vortex colleague in America

Leonard Bakker	Managing Consultation
Richard Bender	Urban Planning, Urban Design
Anne Howerton	Landscape Architecture, Urban Design
David Howerton	Landscape Architecture, Urban Design, Architecture
Don Ichino	Architecture
Ignacio San Martin	Ecology, Geology, Urban Planning, Urban Design, Landscape Architecture, Architecture
Barry Yanchyshin	Landscape Architecture, Urban Design

Vortex colleague in Japan

Satoshi Asakawa	Photograph
Mitsumasa Fujitsuka	Photograph
Yutaka Hikosaka	Architecture and Environmental Design
Shinji Hoshino	Landscape Design and Planning Design
Akito Ide	Communication Design
Shinsaku Imura	Creation
Nobuhiko Inoue	Structural Design
Tatsuo Ito	Lighting Design
Hirokazu Kobayashi	SPREAD, Art Direction and Design
Yoshiko Ninomiya	Harp, Color Design
Mayumi Nishizaki	Landscape Architecture and Architecture
Kazuaki Seki	History of Architecture
Takuro Shimizu	Landscape Architecture
Tokuo Suzuki	Water System Design
Atsunori Takahashi	Landscape Architecture
Yoshihiro Yabe	Landscape Architecture
Haruna Yamada	SPREAD, Art Direction and Design
Hiroshi Yanagihara	mindscape, Landscape Architecture, Urban Design

(Working Group: Graduated from Nagaoka Institute of Design, Ueyama Studio & Yanagihara Studio)

(alphabetical order)

アメリカの渦なかま

レニー・バッカー	マネージング・コンサルタント
リチャード・ベンダー	アーバンプランニング、アーバンデザイン
アン・ハウエルトン	ランドスケープアーキテクチャー、アーバンデザイン
デイビッド・ハウエルトン	ランドスケープアーキテクチャー、アーバンデザイン、建築
ドン・イチノ	建築
イグナシオ・サン・マーティン	生態学、地質学、都市計画、アーバンデザイン、ランドスケープアーキテクチャー、建築
バリ・ヤンチシン	ランドスケープアーキテクチャー、アーバンデザイン

日本の渦なかま

浅川 敏	写真
藤塚光政	写真
彦坂 裕	建築・環境デザイン
星野新治	ランドスケープデザイン、企画デザイン
井出秋人	コミュニケーションデザイン
井村晋作	クリエーション
井上允彦	構造デザイン
伊藤達男	照明デザイン
小林弘和	SPREAD、アートディレクション、デザイン
二宮吉子	ハープ、カラーデザイン
西崎真由美	ランドスケープアーキテクチャー、建築
関 和明	建築史
清水拓郎	ランドスケープアーキテクチャー
鈴木登久雄	水システム設計
高橋宏宗	ランドスケープアーキテクチャー
矢部佳宏	ランドスケープアーキテクチャー
山田春奈	SPREAD、アートディレクション、デザイン
柳原博史	mindscape、ランドスケープアーキテクチャー、アーバンデザイン

(ワーキンググループ：長岡造形大学上山研究室・柳原研究室卒業生)

(abc順)

Exhibition Information

Organizer	Specified Non-profit Corporation for the Inheritance of Architectural Culture (JIA-KIT Architectural Archives) Mitsuru Senda, Kakugyo S. Chiku, Nobuo Mori, Hiroshi Oune, Yutaka Uenami, Junko Sakurai, Yoshihiko Sano, Shinichi Uegaito and Hiroyuki Yano
Co-organizer	Kanazawa Institute of Technology
Cooperation	Ryoko Ueyama + Vortex Colleagues
Sponsor	Nagasaki Prefecture, The Japan Institute of Architects (JIA), Architectural Institute of Japan (AIJ), Japan Landscape Architects Union (JLAU)
Dates	October 19(Sat.) – 26(Sat), 2019
Venue	Grand Hall, Architect's House
Design	SPREAD / Mindscape + Vortex Colleagues
Installation	Waki Process
Staffs	Specified Non-profit Corporation for the Inheritance of Architectural Culture (JIA-KIT Architectural Archives) Koichiro Kanematsu, Takeshi Kirihara, Mikihiro Yamazaki, Kenichiro Shikada, Eichi Sugiyama, Tomoko Taguchi, Jo Toda and Chirudo Sakurai

Sponsors

Mitsui Fudosan Co.,Ltd. / SHIMIZU CORPORATION / KAJIMA CORPORATION / ENVIRONMENT DESIGN INSTITUTE / BellrX CO; LTD. / The MACHINAMI Foundation / TANSEISHA Co.,Ltd. / KENGO KUMA AND ASSOCIATES / Azusa Sekkei Co., Ltd. / KOHYAMA ATELIER / Kenchiku Gahou Inc. / DAIKEN SEKKEI, INC. / Okabe Co.,Ltd. / Kenchiku Shiryo Kenkyusha CO.,LTD. Nikken Gakuin Co.,LTD. / Yasui Architects & Engineers, Inc. / OUNE ARCHITECTS AND ASSOCIATES / KOSO ARCHITECTURAL LABORATORY CO,LTD. / SOGO SHIKAKU CO., LTD. / AXS SATOW INC. / Yamashita Sekkei Inc. / PS Company Ltd. / Architect's House

Photo Credit

FUJITSUKA Mitsumasa p.62,63,64,65,66,67,69,77,78,79(bottom left),84,85
Satoshi Asakawa p.6
Masayoshi Ishii p.79(right)

Catalogue

Prospective Landscapes: A Retrospective – Drawing Exhibition 'Vortex'
Published on September 26, 2019, First Edition

Author and Editor	Ryoko Ueyama / Kazuaki Seki + Vortex Colleagues
Book Design	SPREAD (Hirokazu Kobayashi, Haruna Yamada) / Ryota Iwamatsu
Editorial Cooperation	Kenchiku Gahou Inc.

Published by Specified Non-profit Corporation for the Inheritance of Architectural Culture
Distributed by Kenchiku Gahou Inc.
Daiichi Hayakawaya Building, 2-14-6 Shinjuku, Shinjuku-ku, Tokyo 160-0022
Tel: 03-3356-2568 www.kenchiku-gahou.com
Price: 2500 yen (tax excluded)
Printing and Binding: Taiyo Printing Industry Co.,Ltd.

Any part of this book is prohibited from copying without permission.
©2019 Specified Non-profit Corporation for the Inheritance of Architectural Culture All rights reserved.
Printed in Japan 978-4-909154-63-7

展示概要

主催	NPO法人 建築文化継承機構（JIA-KIT建築アーカイヴス） 仙田 満・竺 覚暁・森 暢郎・大宇根弘司・上浪 寛・櫻井旬子・佐野吉彦・上垣内伸一・矢野裕之
共催	金沢工業大学
協力	上山良子＋渦なかま
後援	長崎県 公益社団法人 日本建築家協会（JIA） 一般社団法人 日本建築学会（AIJ） 一般社団法人 ランドスケープアーキテクト連盟（JLAU）
会期	2019年10月19日（土）— 26日（土）
会場	建築家会館大ホール
デザイン	クリエイティブ・ユニット SPREAD / マインドスケープ＋渦なかま
インスタレーション	脇プロセス
スタッフ	NPO法人 建築文化継承機構（JIA-KIT建築アーカイヴス） 兼松紘一郎・桐原武志・山崎幹泰・鹿田健一朗・杉山英知・田口知子・戸田 穣・櫻井ちるど

協賛

三井不動産株式会社／清水建設株式会社／鹿島建設株式会社／株式会社環境デザイン研究所／株式会社ベルックス／一般財団法人住宅生産振興財団／株式会社丹青社／隈研吾建築都市設計事務所／株式会社梓設計／有限会社香山壽夫建築研究所／株式会社建築報社／株式会社大建設計／株式会社岡部／株式会社建築資料研究社 日建学院／株式会社安井建築設計事務所／株式会社大宇根建築設計事務所／株式会社構想建築設計研究所／株式会社総合資格 総合資格学院／株式会社佐藤総合計画／株式会社山下設計／ピーエス株式会社／株式会社建築家会館

写真

藤塚光政 p.62、63、64、65、66、67、69、77、78、79（左下）、84、85
淺川 敏 p.6
石井雅義 p.79（右）

図録

天と地をつなぐランドスケープ 渦展
2019年9月26日 初版第一刷発行

編・著	上山良子／関 和明＋渦なかま
装丁・本文デザイン	SPREAD（小林弘和・山田春奈）／岩松亮太
編集協力	株式会社 建築画報社
発行	NPO法人 建築文化継承機構
発売	株式会社 建築画報社 〒160-0022 東京都新宿区新宿2-14-6 TEL：03-3356-2568 www.kenchiku-gahou.com
定価	2500円（税別）
印刷・製本	太陽印刷工業 株式会社

落丁・乱丁本はお取り替えいたします。無断で本書の全体または一部の複写・複製をすることを禁じます。
©2019 NPO法人 建築文化継承機構 All rights reserved.
Printed in Japan 978-4-909154-63-7